Darinka Sikošek

Two pages from my methodological note-book: CHEM-TEACH&LEAR

Darinka Sikošek

Two pages from my methodological note-book: CHEM-TEACH&LEAR

How can students get Knowledge-Skills-Attitudes-Responsibility for Chemical Substances-Laboratory Techniques-Equipment?

LAP LAMBERT Academic Publishing

Impressum / Imprint

Bibliografische Information der Deutschen Nationalbibliothek: Die Deutsche Nationalbibliothek verzeichnet diese Publikation in der Deutschen Nationalbibliografie; detaillierte bibliografische Daten sind im Internet über http://dnb.d-nb.de abrufbar.
Alle in diesem Buch genannten Marken und Produktnamen unterliegen warenzeichen-, marken- oder patentrechtlichem Schutz bzw. sind Warenzeichen oder eingetragene Warenzeichen der jeweiligen Inhaber. Die Wiedergabe von Marken, Produktnamen, Gebrauchsnamen, Handelsnamen, Warenbezeichnungen u.s.w. in diesem Werk berechtigt auch ohne besondere Kennzeichnung nicht zu der Annahme, dass solche Namen im Sinne der Warenzeichen- und Markenschutzgesetzgebung als frei zu betrachten wären und daher von jedermann benutzt werden dürften.

Bibliographic information published by the Deutsche Nationalbibliothek: The Deutsche Nationalbibliothek lists this publication in the Deutsche Nationalbibliografie; detailed bibliographic data are available in the Internet at http://dnb.d-nb.de.
Any brand names and product names mentioned in this book are subject to trademark, brand or patent protection and are trademarks or registered trademarks of their respective holders. The use of brand names, product names, common names, trade names, product descriptions etc. even without a particular marking in this works is in no way to be construed to mean that such names may be regarded as unrestricted in respect of trademark and brand protection legislation and could thus be used by anyone.

Coverbild / Cover image: www.ingimage.com

Verlag / Publisher:
LAP LAMBERT Academic Publishing
ist ein Imprint der / is a trademark of
OmniScriptum GmbH & Co. KG
Heinrich-Böcking-Str. 6-8, 66121 Saarbrücken, Deutschland / Germany
Email: info@lap-publishing.com

Herstellung: siehe letzte Seite /
Printed at: see last page
ISBN: 978-3-659-57843-4

Competence is our passport for tomorrow,

to the people who prepare for it today.

I dedicate this book to my:

    Students-prospective teachers of chemistry,

      Graduates of chemical education and

        Former secondary school pupils ...

Contents

Preface

This publication on methodological approaches about chemical education, titled *Two Pages from my methodological note-book: Chem-Teach&Lear* has been prepared to provided some issues of the many opened questions about effective teaching and learning natural sciences such as subjects Sciences and Chemistry for Secondary Vocational Education that arise in teacher's everyday preparation on pupils / students learning activities. These activities concerned with the learners' competences, such as the knowledge, skills, relations, attitudes and responsibility, needed for safe handling of chemicals and equipment in a chemistry laboratory. Thus, the strategy competence-oriented activities provide the student the opportunity to obtain not only topical subject-specific competences, but also a number of for consumers (also known as object-independent competences that an individual is realized during the provision of targeted actions of their future professional work).

The volume consists of two booklets: (1) How to use the chemical concept "Bicarbonate of Soda" for teaching and learning chemistry and science; (2) What is hidden in pupil / student Laboratory Notebook.

First topic deals with the chemical concept soda as the usual term which using can be represented by the inter-subject planning and connecting for the tendency to acquire higher level of competences' acquiring (achieving the desired learning outcomes).

The central subject of the second topic is pupil / student as an evolving experimenter, because experimentation is a basic method of learning chemistry as well as chemical research.

It is well known that teacher neither "gives" any of competences to his/her learner, nor it cannot be "accepted" by him/her, but it should be acquired by every-learner with his /her own mental and experimental (training) activities.

The methodology of designing didactic teaching material which, given a particular set of minor differences for instruction worksheets. The Learning-Worksheet (Topic 1) includes the following instructional parameters: problem, individual tasks, general and specific operational instructions for implementation of the activities defined roles in relation to the planned practiced methods of work. In other type of didactic materials, i.e. the learning-worksheets (topic 2) are included the following parameters: (1) introductory note, (2) self-asked general questions, supplemented with answers- i.e. operative instructions, followed by two concrete tasks: the first relates to the defined laboratory work and other on the preparation of a pupil's / student's personal catalogue of chemicals throughout the school year.

Maribor, June 2014

Darinka Sikošek

Chair of Didactics of Chemistry, Faculty of Natural Sciences and Mathematics, University of Maribor, Slovenia

Topic one

I   How to use the chemical concept "Bicarbonate of soda" for teaching and learning chemistry and science

Motto: A teacher must prepare a learner to correct himself;
Otherwise, the result of learning is a skill depending on the teacher's presence
(author)

Summary

This booklet comprises didactic material dealing with soda as one of the daily means used for leaving and raising dough. This learning content as a part of the basic chemical educational standard is also an opportunity to carry out activities during which students acquire abilities necessary for the implementation of the tasks and needs of both professional work and daily life.

The theoretical section presents the definition of this curricular concept (included in the Catalogue of knowledge on the subject of Natural Science, 132 hours, Secondary Vocational Education Program with aspects of the objectives, content and activities.

Specifically defined are the topical generic competences as well as individual scientific competences which can be achieved by learning through the use of this material. The competence aspects of learning, based on current background knowledge, are particularly highlighted in the learning guide for students.

The practical part of this book presents material that is methodologically treated using the method of experimental work and heterogeneous group work. The proposed activities are taken into consideration the didactic principle of differentiation, according to stakeholder capabilities, the principle of individuation.

In addition to the solution of tasks and evaluation of the completion of practical works student competence achievement includes a topical dictionary of chemical terms and a list of recommended reading for students, as well as additional literature for teachers.

Historians, Experimenters and Consumers in our ChemLab (Photo: D. Sikošek)

Introduction

In the frame of Slovene Secondary Vocational Educational Programmes such as food-processing worker–confectioner, Science in connection with The Fundamentals of Biotechnology with Nutrition (as one vocational subject) is defined as a topical subject. Within the curricular concepts of the teaching unit on Water Solutions, there occurs the well-known concept of BiCarbonate of Soda. Among the important curricular goals of this didactic unit are the following: (1) study of the properties of aqueous solutions (e.g. pH) deriving from self-supported experimental observation; (2) evaluation of the meaning of aqueous solutions for living and working; (3) knowledge of various food supplements and explanation of their presence. The author of present contribution have defined the potential achievement of these goals considering aspects of the topical generic and natural competences. Deriving from present competence's conscription, the necessary student activities are planned, taking into account the didactic principles; the choice of optimal methods and social forms follows.

The didactic material "BiCarbonate of Soda (in short called soda) contains both theoretical and applied sections. The theoretical base includes a basic presentation of the means for dough raising as a curricular concept, as well as the principle of differentiation and individualisation as a necessary didactic point. However, the applied section also contains the material needed by both partners in the learning process: the teacher and the student. Instructions in the form of a didactic proposal and the full performed directions for completion are included to fulfil the teacher's needs. Besides these, a learning guide and some learning / experimenting sheets have been prepared to help in the performance of planned student activities.

The other components of this didactic material includes topical connections to the other teaching / learning units, presents pre-knowledge with appropriate learning sources and, of a course, a vocabulary of new chemical and didactic terms.

A topical evaluation instrument has been created for the teacher-evaluator.

Theoretical section

1. The teaching and learning approach: **Program's purpose, objectives, activities and methods of the learning unit**

<u>Program</u>: **Secondary Vocational Education (SVE)-***foodworker (*confectioner*)*
<u>Subject</u>: **Sciences** in inter-subject connection with the subject **Base of Biotechnology for Nutrition**
<u>Learning unit</u>: **Aqueous solutions → Soda** (as curricular concept)
<u>Guidelines goals and activities</u>: See the operational target-activities fragment from the Catalogue of knowledge on two subjects: Sciences 132 lessons) and Base of biotechnology for nutrition (Table 1) and taxonomic levels of target activities for the content section "Soda" (Table 2).

Table 1: From curricular learning objectives across activities to outcomes

Curricular content section: **Aqueous solutions**	
Learning Objectives (topical section)	**Examples of activities, *methods***
✓ observation in a separate experimental study of the properties of aqueous solutions (pH, or acidity and alkalinity, electrical conductivity); ✓ evaluation the significance of aqueous solutions for life and work: ✓ familiarity with different types of additive; ✓ explanation of the importance of various food additives;	Measuring pH values of aqueous solutions of acids and bases encountered in everyday life and professions; Complying with the provisions for the use of additives in food. ***Experiments:*** *- demonstrate the properties of carbon dioxide;* *- evidence of carbon dioxide using calcium hydroxide;* *- heating sodium bicarbonate;* *- testing the presence of carbon dioxide by burning cod.*

Anticipated curricular outcomes ➡ Students can:
✓ distinguish between acidic and basic solutions, and use pH value to assess the strength of acids / alkalis;
✓ provide examples of the use of acids, bases and salts in their daily lives;

Table **2**: Example of content correlation with the **Soda** section

Target activities (*taxonomic level*)
Contextual password set : **History of soda** ✓ list the methods of leaving and raising dough *(knowledge / Bloom);* ✓ practice writing of chemical reactions *(knowledge / Bloom)*; ✓ eliminated useful information on working with text *(work resource / Marzano).*
Contextual password set: **Soda for experimental enthusiasts** ✓ list the methods of leavening and raising dough *(knowledge / Bloom);* ✓ write chemical formulas *(understanding/Bloom);* ✓ provide properties of matter *(understanding / Bloom);* ✓plan and carry out an experiment *(application / Bloom);*
Contextual password set: **Soda in our lives** ✓ use a various electronic and book resources *(working with resources /Marzano);* ✓.carry out research *(understanding / Bloom);* ✓.identify everyday products containing soda *(application / Bloom).*

2. The teaching and learning approach: **The Competences Activities**

Competences are defined as a combination of knowledge, skills and relations according to circumstances. Following DESECO (2002), we distinguish the following competences: generic, particular and subject specific. However, the generic competences are defined as the ability to perform various activities such as, collection of information, literature analysis, organisation and interpretation of information as well as synthesis of conclusions (prepared by Mayer, 1991).The optimal generic competence acquired by teaching and learning using this didactic material on soda is the *ability to learn and to solve problems*, the operation of which is presented in Table 3.

Table **3**: Operation of the competence **"learning and problem solving ability"**

Student's knowledge	Student's skills	Student's relations (attitudes)	Teacher's notes
✓ Knowledge of strategies to solve problems;	✓ Planning their own learning; ✓ Designing and implementing plans; ✓ Implementation of activities.	✓ Responsibility for the effective implementation of activities; ✓ Quest for self-development; ✓ Commitment to achieving individual and common goals.	

Other generic competences, developed by the students include the following abilities:

(1) the collection of information, (2) synthesis of conclusions, (3) working independently and in a team, as well as further effectiveness in, (4) planning and organising of work and (5) verbal and written communication.

Of course, the implementation of the contextual password Soda is linked with the performance of science competences (Table 4), which are perceived as the intersection of knowledge, skills and attitudes in the fields of Chemistry, Biology and Physics (Špernjak & Šorgo, 2009).

Table **4**: From contextual password sets to their **target competences**

Contextual passwords set	Competence targets (aims)
(1) **History of soda**	✓ awareness of the importance of real chemical processes and the connection process with the characteristics (properties) of the products we use in everyday life;
(2) **Soda for experimental enthusiasts**	✓ awareness of possible links between the chemical processes industry and the experimental work in the school chemistry laboratory;
(3) **Soda in our lives**	✓ ability to use chemical knowledge and terminology in everyday life as a consumer;

3. The teaching and learning approach: **Considering Didactic Principles**

Besides the competence aspect the other important didactic focus is the consideration of direct realisation of the internal goals-activities of the principle of differentiation and individualization. Presented didactic material represents the instance (in terms of didactic theory) direct implementation of target-activity starting points of internal differentiation and individualization.

In general, the achieving this demands a variety of learning process parameters, especially competence goals, activities, content, methods and social forms. Deriving from theoretical origins three forms of differentiated and individualized learning work (activities) are planned (Ošlovnik, 2004):

(1) *Self-differentiated group work*: students suggested the optimal group tasks, taking into account the abilities of their members (self-evaluation);

(2) *Self-differentiated individual work:* in this form of individual work students solve or undertake various challenging tasks with similar content, where each student self-estimates and determines how many tasks can be performed/solved by himself / herself and to which taxonomic levels;

(3) *Individualised group work* this takes place with a number of limited groups, as independent individual learner activities-individual tasks, operating as components of whole group's task.

Implementation of internal individualization and differentiation, after Ošlovnik (2004), shows in such a teaching configuration that engages all students, regardless of differences in their perception of the subject matter. To do this, the teacher must consistently take into account all the parameters of the learning process, so that he/she varies the objectives, activities, content, methods, verification, etc.

4. **The teaching and learning approach:** Occupations and Tasks for this Method

For the realization of these competence activities among our learner-confectioners, the performance of whom would be suitably differentiated and individualised, we have chosen the role play method, with experimentally supported problem approach as the optimal method of learning. The learning content (so-called problem) is defined thus: Soda or sodium carbonate appears as anhidricum in formula Na_2CO_3, or in crystal form as the formula Na_2CO_3 x 10 H_2O. It is indispensable because of its exceptionally wide usage. To food-handlers (especially confectioners) as well as to all of us, bicarbonate of soda, the formula $NaHCO_3$ is familiar from our home kitchen. Let us collect more information about this important chemical, which is also as well as a component of popular sweets!

This problem is structured into the three following items:
(1) The history of soda; (2) Soda for experimental enthusiasts; (3) Soda in our life.

What followed was the formation of three role play groups: historians, experimenters and consumers. Optimal competences attainable by the performance of these roles mentioned roles are clearly presented in Table **1**.

Table **5**: Description of the competences of the **role play** groups

Generic **Competences :** $\checkmark$ability to ...; *becoming aware of ...* Legend: **Historians, Experimenters, Consumers**	H	E	C
$\checkmark$learning & problem solving;	$\checkmark$	$\checkmark$	$\checkmark$
$\checkmark$ collecting information;	$\checkmark$	$\checkmark$	$\checkmark$
$\checkmark$safe experimenting /chemicals lab technique, protection;	/	$\checkmark$	
$\checkmark$self-support & team work;		$\checkmark$	$\checkmark$
meaning real processes in the chemical industry regarding the properties of everyday products;	*		
meaning the possible connections of chemical processes with the school laboratory;		*	
using chemical knowledge and terminology in everyday life as a consumer;			*

12

Didactic material that has been suitably adapted to the problem is required for this role-play method, supported by experimenting and carrying out in an individual social form.

Our pastry-cooks who are playing **historians** make use of both recipes and textbooks or handbooks as sources of information. The three individual tasks are defined as follows: the first individual task requires a careful review of the recipe in terms of useful methods for raising the dough; the second task is two-fold: first (a) it requires making notes about the procedure for soda production; second (b) it requires us to write down the full chemical equations for the method of soda bicarbonate production; the final (third) individual task is to make note the correct explanation for the meaning of raising and leavening agents in confectionery.

Our pastry-cooks who are playing the role of **experimenters** they plan and carry out the experimental work using *individualised group teamwork* from a pre-designed personal Implementation Plan.

The first task for the individual representing the coordinated participation of the group of "historians" is to obtain information about resources for raising dough.

The second task of the two experimenters, defined as the selection of the current experiment include: (a) study of the theoretical basis and (b) experimental pre-treatment (testing of laboratory methods and procedures).

The third task, also two-fold is defined as the demonstration of two experimental passwords: "Simple honey dough" and "Apricot cake".

Our pastry-cooks who are playing the role of **consumers** carried out the following individual tasks in the context of substantive password "Soda in our lives":

(1) Explanation of how substances with a similar (even identical) chemical composition (e.g. soda) are used for different purposes;

(2) Acting as consumers who know how to use chemical knowledge and terminology for: (a) industrial use (to present soda as an important industrial raw material with diverse uses), and (b) for the individual needs of everyday life.

5. Theoretical chemical content section "soda"

The main focus of this sub-section is on the curricular concept, called **Raising Agents** and the didactic parameter, called **Principle of differentiation and individualization.**

The raising means (agents) added to the dough cause the formation of gas (bubbles of carbon dioxide), which renders the bread and pastry light. For this purpose sodium bicarbonate is commonly used, either alone or in combination with tartaric acid.

The carbonates and hydrogen carbonates in reaction with acid release carbon dioxide, forming salt and water, which can be written as the simplified reaction:

$$\text{dough} \rightarrow NaHCO_3 + H^+ \longrightarrow \text{dough}: CO_2 \text{ (g)} + H_2O$$

In terms of didactic theory, the given didactic material represents one sample of the direct implementation of the basic origins of internal target-activity differentiation and individualization. The implementation of interior individualization and differentiation (after Ošlovnik, 2004) shows in a teaching configuration that engages all learners (students), irrespective of differences in their perception of the subject matter. However, in order to succeed, the teacher must consistently take into account all the parameters of the learning process, such as the varying objectives, activities, content, methods, verification approaches and so on.

In the classroom it is possible to achieve (referring to the above-mentioned source) the four forms of **differentiated and individualized learning work** (activities):

(1) *Self-Differentiated Teamwork*-within a group the students themselves determine their task and activities regardless of their abilities (self-evaluation).

(2) *Differentiated Individual work*-represents a form of individual work in which students solve or undertake various challenging tasks with similar content.

(3) *Self-Differentiated Individual work* - a form of individual work, in which students solve or undertake equally difficult tasks with the same content. How many tasks and to what degree of difficulty will be those solved, evaluated and determined individually.

(4) *Individualized Teamwork*-performed as self-individualised work by students in numerically limited groups on individual tasks that are an integral part of the group assignment.

6. Links to other teaching units

Knowledge acquired through this teaching unit offers an inter-connection with the subject "Fundamentals of Food Biotechnology", especially the content set "Food Additives". The emphasis is on learning about different nutritional supplements and comparing their chemical composition.

7. Previous knowledge

Current knowledge representing other concepts acquired from the set of "Aqueous Solutions and "Food Additives" from the Catalogue of knowledge.

8. Glossary of chemical terms

Soda, sodium carbonate (the formula Na_2CO_3.) is a crystalline, highly water-soluble substance; aqueous solutions strongly alkaline. Sodium carbonate in the chemical industry is extremely important raw material (production of soap, various types of glass, detergents and water softeners).

Sodium bicarbonate (sodium hydrogen carbonate), baking soda, the formula $NaHCO_3$) is a crystalline not the best water-soluble substance. It is produced by the introduction of carbon dioxide into a cooled, saturated sodium carbonate solution, making the solution slightly alkaline. Therefore, in technical chemistry, Sodium Carbonate is commonly used to neutralize acid (in the past, for neutralizing stomach acid).

Ammonia (the formula NH_3 is smelly, highly water soluble gas, aqueous solution acts as a weak base. It is one of the most important substances (raw-material) in the chemical industry. This compound is important in the industrial production of sodium carbonate after Solvay process.

Carbon dioxide (the formula CO_2) gas is colourless, odourless and tasteless; this readily water-soluble solution reacts slightly acidic because 1% of the gas in the water solution is bound in carbonic acid. The carbon dioxide content in air varies slightly, depending on the intensity of photosynthesis taking place in plant organisms.

9. Learning resources

Recommended

Atkins, P.W., Clugston, M.J., Frazer, M.J., Jones, R.A.Y. (1988). CHEMISTRY-Principles and Applications, illustration P.W.Atkins. Longman Group UK, London.
Campbell, H. Bonnie, Ruptic, Cynthia (1994). Practical Aspects of authentic assessment:
Putting the pieces together. Christopher Gordon Publishers.
Chemistry: The Salters' Approach Pupil's Activity Book. University of York Science Education Group 1987, 1988. Heinemann Educational Publishers Great Britain.
Duschner, A., Kunze, W., Moosburger, G., Nagel, H., Schlegel, R., Wisniewski, H. (1989) Chemie fur bayerische Realschulen, Jahrgangstufe 9. In: Natur und Technik. Cornelsen Verlag, Berlin.
Hill, G., Holman, J., Lazonby, J.,Raffan, J., Waddington, D. (1990). Chemistry Group. Heinemann Educational Publishers Great Britain.
Schülerduden, Chemie, Bibliographisches Institut & F.A.Brockhaus AG. Mannheim, 1995, 2001.
The *Solvay process* or ammonia-soda process, available 30.10.2013, http://en.wikipedia.org/wiki/Solvay_process,

Additional

Skvarč, M. (2004). From design to testing and assessment knowledge of chemistry in elementary school. (reviewer D. Sikošek). In: Towards a New Culture of Teaching. The National Education Institute of the Republic of Slovenia. Ljubljana. (only in Slovenian).
Furman, V. (D. Sikošek- mentor) (2003). Authentic tasks such as evaluation material in process verification and assessment of students' knowledge in elementary chemistry education. Short thesis. Department of Chemistry, Faculty of Education. University of Maribor. (only in Slovenian)
Ošlovnik, S. (D. Sikošek- mentor) (2004). Internal differentiation and individualization of curriculum reform of teaching chemistry in elementary school.

Thesis. Department of Chemistry, Faculty of Education, University of Maribor. (only in Slovenian)

Lekše, I. (1999). Kakšne naj bodo domače naloge? School counselling, Volume 4, No. 2/4,p. 22-31. ? The National Education Institute of the Republic of Slovenia .Ljubljana. (only in Slovenian);

Ilc, Marija-sestra Vendelina (1995). Pecivo sestre Vendeline. Založba Vale-Novak. Ljubljana. p. 105, p. 31- 33. (only in Slovenian)

View from the Museum of Berzelius in Stockholm
(Information Department of the Royal Swedish Academy of Sciences)

Applied section

The material for practical work is extensive and includes the following components:

(1) teaching guide for teachers referring to the planning (i.e. didactic proposal), A1, and implementation guidelines (instructions), A2;

(2) instructions for students (i.e. teaching guide for students), B;

(3) teaching materials in the form of learning and worksheets for direct implementation of student practice in relation to their assumed methods' roles,
C: ✓ group learning-sheet 1 with individual problem tasks for the "historians',
   ✓ group worksheet 2 with individual problem tasks for the Experimenters and the consolidated,
   ✓ group learning sheet 3 with single individual assignments, separate for each consumer;

(4) solutions of the individual problem tasks, D, for substantive passwords 1,2 and 3.

A. Instructions for teachers

1. Didactic proposal

Preliminary remarks:
Current **didactic principles** that can be applied for all content keywords are the following: the principle of differentiation and individualization the principle of gradualness, the principle of actuality and the principle of vitality.

Table **6**: Didactic proposal for suggested **learning methods and social forms** concerning the activities carried out

Function tasks completed	Student activities	Learning Methods	Social Learning Forms
Contextual keyword **1: History of soda**			
✓ consolidation, ✓ dissemination of comprehended content;	✓ work with sources of chemical and historical areas;	✓ work with text;	✓ individualized group work (teamwork);
Contextual keyword **2: Soda for experimental enthusiasts**			
✓ consolidation, ✓ deepening comprehended content;	✓ planning and implementation of experimental work, ✓ work with text sources;	✓ experimental work of students, ✓ work with text;	✓ Self-differentiated group work;
Contextual keyword **3: Soda in our lives**			
✓ dissemination of comprehended content, ✓ motivation;	✓ work with electronic and book resources, ✓ independent research students;	✓ problem solving;	✓ Self-differentiated individual work within the common task;

2. Implementation guidelines

1. In the predictions, the teacher presents three possible approaches to learning. In each of these approaches, students will discover one of the versatile products of chemical technology: soda (especially sodium bicarbonate). Students should be classified as follows: the first group comprise historians, the second the experimenters, and the third the consumers.

2. The teacher distributes worksheets with instructions to each group. Students write down the solutions in their notebooks.

3. Completion of the work of each group takes place as preparation for the team report (which shows all the solutions to specific tasks for which the members of the particular group are responsible), which the teacher has also reviewed and evaluated. The plenary session also includes student inter-group presentation reports.

4. Teachers allow students access to literature or the Internet, and help with the practical execution of the experimental work of demonstrating carbon dioxide.

5. Description of strategy: after the teacher's presentation of the selected topic, follows the formation of groups and the distribution of learning material and instructions, which read as follows: it is necessary to independently prepare a plan for the implementation of individual tasks.

An integral part of the organization's work is an implementation plan for the individual work of each member of each selected group. A scheme for the implementation plan is given to each student; a completed spreadsheet for all activities completed (in accordance with the implementation plan) is submitted by each student (after completion) to the teacher for checking, but it is also helpful to him/her in the evaluation of the generic competences exhibited by each student.

Work on the content set "Soda" should take place continuously over a fourteen-day period, so that most of the activities are completed outside school hours. For coordinating workflow implementation plans, there must be complete cooperation among all members of the group carrying out the activities of each group (range, 10-minute lesson). In the next three hours, the teacher devotes only ten minutes for group meetings; other work must be carried by students at home or in the laboratory and library. The fourth hour is dedicated entirely to presentation and evaluation of the students' work.

Table **7**: Outline of the implementation plan for **individual work** (a personal record of focus for each (thematic) group member)

Thematic group:	*Student (learner) as performer:*	
Activities carried out	**Terms**	**Notes**

B. Learning guide for students

The tasks are performed in order to introduce students to research work associated with seeking varied sources of information and selecting the most relevant information, which in this case involves the best method for leavening dough.

First, particular emphasis is placed on the use of chemical knowledge for vocational purposes; this involves implementing the planned activities related to student awareness of the connection between chemical reactions and processes and the ingredients, preparation and manufacturing process of bakery products.

Another aspect of implementing these tasks is to learn and clarify the possible diversity of one or more related substances (products) for vocational (industrial use), or daily domestic needs.

C. Didactic material: **Learning / Working Sheets**

Group 1 Historians   **group learning sheet 1**

Contextual keyword 1:

21

Problem

Soda or sodium carbonate acting as an anhydrous substance in the formula Na_2CO_3 or in the crystalline form of formula $Na_2CO_3 \times 10\ H_2O$ is indispensable because of its diverse utility.

Food experts (especially confectioners), as well as all of us in the kitchen are familiar with the third form of soda for baking, so-called sodium bicarbonate of the formula $NaHCO_3$. So, let us discover more about this important chemical as a component in popular cakes.

Chocolate muffins
200 g of flour, 100 g sugar, approx. 70 g chocolate,
2 eggs, 0.15 L milk 60 g butter, 2 tablespoons cocoa
powder, 10 g vanilla sugar, 1/2 teaspoon, baking
soda, a pinch of salt;

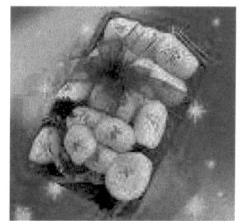

Honey confetti
750 g flour, 500 g honey, 0.1 L of water, 0.1 L kir liquor,
10 g ammonia (dissolved in water), 150 g citroma
(magnesium citrate), 1 egg;

Implementation guide for historians
The group task for this contextual keyword is divided into three specific tasks, distributed among the individual members of the group by the manager. The first task requires the learner's knowledge of the methods for raising and leavening, the second involves the use and understanding of the work of literature; in the third it is necessary to use written information to connect and explain by chemical facts.

Recommended literature
Chemistry - Principles and Applications / P. W. Atkins, M.J. Clugston, M.J. Frazer, R.A.Y. Jones/,Longman Group UK, London, 1988.

Individual tasks

1. Which methods for leavening and raising and dough you recognize in the recipes above? *(Review.)*

2a. The discoverer of the process of producing soda became wealthy for the ingenuity required to complete it, and especially owing to the great importance of this product. Identify what the process going. *(Write down the relevant information for this process.)*

2b. How runs a chemical reaction based on the process of producing soda? *(Write the full equation for this reaction.)*

3. What is the significance of the identified methods in favour of rise and loosening for their use in confectionery industry? *(Write down your finding.)*

Group 2 Experimenters   **Group learning sheet 2**

Contextual keyword 2:

Soda for experimental enthusiasts

Problem

We remind you of your last birthday promise to classmates that from now on pastries will no longer be purchased from the confectionery (or bakery), but will be prepared by you at home. So, you will become amateur (and later perhaps also professional) confectioners, which means you will be experimenting with edibles: preparing raw materials (flour, dough and toppings), creating (by hand or machine) products, baking them and inserting fillings.

As it was said done to be so well done, therefore successful at work!
or using Latin folk saying: Detto fatto!

23

Implementation guide for experimenters

Group task of this contextual keyword is divided into three specific tasks, and each member of the group can choose a task at their own discretion, which the group head records (this role is taken by one of the experimenters).

The first task was designed to identify methods for leavening and raising, the second for seeking information in the literature and the hypothesis, while the third requires the planning and realization of the experiment and the proper evaluation of obtained results.

Apricot cake
3 eggs, 13 g sugar, 220 g flour, 1 baking powder,
150 mL milk, 1 tablespoon rum, 9 g butter,
1 cup of apricot filed;

Easy honey cakes
200-300 mL milk, 400 g powdered sugar,
fragrance of sweet cakes
1 egg, 1 kg of wholegrain flour, 1 teaspoon of baking soda;

Individual tasks

1 Which means for loosening and rising recognize from the above recipes? *(Read and print.)*

2a. What's the chemical formula of the extracted resources?
(Find them in the literature, and write.)

2b. Determine which in the written chemical formulas the joint compound is that has the ability to stretch and how it might be obtained by appropriate reaction.*(Compare the formula of the compounds and read out from them the right.)*

3. Experimentally verified the formation of the substance, which allows loosening and leavening of dough. *(Carry out the experiment.)*

Contextual keyword 3:

Problem

Polona the confectioner comes to work and explains to her colleagues:

"You know Majda, nothing is clear to me anymore. Yesterday I baked honey cakes, which as you know require soda, but today I bought a dishwashing detergent with effect of soda."

"Really, just what effect it has the means of loosening dough if it is used for washing dishes?", Majda thinking aloud.

"And that's not all, a pharmacist at the pharmacy explained to my husband that his pills for controlling stomach acid also contain some kind of soda."

"Polona, I don't know how to explain it", says Majda.

Polona thoughtfully adds, *"But I would still like to understand what's going on."*

Individual task mine, yours and ours:

We are all consumers, so let's explore this confusion on the part of Polona the consumer and so help her to understand "what is going on."

Implementation guide for the consumers

1 Each member of the consumer group chooses their sources of information (choose between different electronic and book resources, come in to the pharmacy and talk to your pharmacist); all this information must also be recorded by the head of the group.

2. After completing a query, submit a written explanation to the head of the consumer group. So, both of you can together (with the teacher) transmit it to Polona.

Contextual keyword 1:

Solutions to individual tasks:

1. ammonia, sodium bicarbonate;

2a. The Solvay process (ammonia-soda process) for the industrial production of sodium carbonate (soda), which was developed by the Belgian chemist Ernest Solvay (1838-1922). In this procedure, the different solubility of the compounds involved in the process is skilfully utilized.

2b. Carbon dioxide is introduced into ammoniated salt-water:

$$CO_2 \text{ }_{(g)} + NH_3 \text{ }_{(aq)} + NaCl \text{ }_{(aq)} + H_2O \text{ }_{(l)} \longrightarrow NaHCO_3 \text{ }_{(s)} + NH_4Cl \text{ }_{(aq)}$$

The resulting sodium hydrogen carbonate is insoluble in these conditions; first it is filtered and then it is heated in rotary calcinators:

$$2NaHOCO_2 \text{ }_{(s)} \longrightarrow Na_2CO_3 \text{ }_{(s)} + H_2O \text{ }_{(g)} + CO_2 \text{ }_{(g)}$$

In summary:
The simplified chemical reaction of the overall four phase cyclic process is:

$$2 NaCl + CaCO_3 \rightarrow Na_2CO_3 + CaCl_2$$

The chemical phases reactions are:

Phase 1:

$$\underline{\mathbf{CaCO_3}}(s) \rightarrow \underline{\mathbf{CO_2}}(g) + \underline{CaO}(s)$$

Phase 2:

$$2 \underline{NH_4Cl}(aq) + \underline{CaO}(s) \rightarrow 2 \underline{\mathbf{NH_3}}(g) + \underline{CaCl_2}(aq) + \underline{H_2O}(l)$$

Phase 3:

$$\underline{\mathbf{CO_2}}(g) + \underline{\mathbf{NaCl}}(aq)\,1 + \underline{\mathbf{NH_3}}(g) + \underline{H_2O}(l) \rightarrow \underline{\mathbf{NaHCO_3}}(aq) + \underline{NH_4Cl}(aq)$$

Phase 4:

$$2 \underline{NaHCO_3}(aq) \rightarrow \underline{\mathbf{Na_2CO_3}}(s) + \underline{H_2O}(l) + \underline{\mathbf{CO_2}}(g)$$

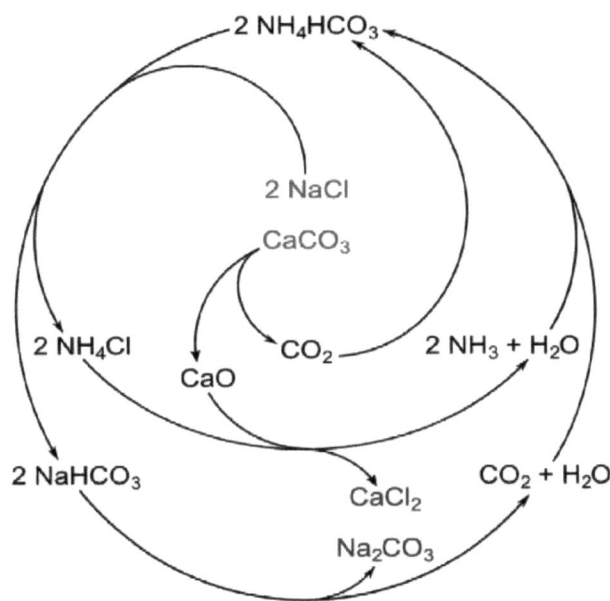

Colour legend: green = reactants, black = intermediates, red = products

27

Contextual keyword 2:

Soda for experimental enthusiasts

Solutions to individual tasks:

1. Sodium bicarbonate, Baking powder;

2 a. Sodium bicarbonate and baking powder containing sodium bicarbonate, which when heated decompose, releasing carbon dioxide. In addition, baking powder also contains a weak acid, such as 2,3-dihydrobutanedioic acid (tartaric acid). In the reaction carbon dioxide is released, causing the dough to rise.

2b. The reaction of decomposition of sodium hydrogencarbonate (bicarbonate of soda):

$$2NaHCO_{3(s)} \xrightarrow{\text{heat}} Na_2CO_{3s)} + H_2O_{(g)} + CO_{2\,(g)}$$

3. Possible evidence regarding to selection of the experiment:

a) the saturated lime water is opaque because of bubbled CO_2 fine white precipitate forms;

b) the smoulder cod is extinguished in the presence of CO_2.

Contextual keyword 3:

Soda in our lives

Solution to consumers task (explanation for Polona):

➡ Washing soda, which is a crystalline form of sodium carbonate decahydrate (the formula Na_2CO_3 x 10 H_2O) is used as a cleanser and water softener used.

➧ Baking soda (sodium hydrogen carbonate is used as an antacid for alleviating the effects of "stomach acid" (for heartburn).

Since this salt, as well as other similar salts (e.g. potassium bicarbonate, sodium phosphate, calcium phosphate) is absorbed into the bloodstream, this can cause metabolic alkalosis; therefore, their use is no longer recommended.

Antacids are a commercially available group of drugs that can reduce acidity of gastric juice pH 2 to pH 3-4, which lessens the symptoms of stomach problems.

E. Evaluation instrumentation

This unit comprises an evaluation sheet for evaluating the level of achievement in the fundamental components (knowledge, skills and attitudes) of the four types of generic skills (information gathering, learning and problem solving, gathering information and learning, independent work and teamwork) that students acquire by performing of various activities related to the three roles: historians, experimenters and consumers.

A descriptive evaluation of the level of knowledge, skills and attitudes is proposed, with regard to the three-point scale: inadequate, satisfactory, commend able.

Universal generic competency: this competency is defined as common to all the tasks carried out by students of both groups, the historians and the experimenters and proposed the following scale (degree) for evaluation of the given competence's activities: Inadequate (I), Good (G), Commendable (C).

Table 9: Descriptive evaluation of universal **generic competences** common to historians and experimenters

Generic competency: **The ability to work independently and in a team**
✓ degree of organization and functioning of those responsible for individual tasks planned in the framework of the group roles; **I G C** ✓to the members of the same / other groups, the person responsible for task can clarify the objectives of a specific task and explain how to solve this part of the joint group task; **I G C**
Teacher comment:

Specific generic competences: these are defined within the individual tasks of roles for groups 1, 2 and 3, wherein all the competency activities are evaluated using the above descriptive scale (degree): Inadequate (I), Good (G), Commendable (C).

Contextual keyword 1:

Group 1: Historians

Table 10: Descriptive evaluation of key generic competences for **historians**

Generic competency : The ability to **collect information**		
student's **Activities / individual Tasks**	student's **Knowledge**	student's **Skills**
1.*Identification* **methods** *for raising and leaving dough;*	He/She knows the definition of the information given; **I G C**	✓ He/She can obtain information from different sources; **I G C** ✓ He/She has a critical attitude towards the information collected; **I G C**

Teacher comment:		
2a. Description of the **Solvay process** *for the production of soda;*	He / She is acquainted with current knowledge; **I G C**	He / She has the necessary reading / communication ability; **I G C**
Teacher comment:		
2b. Recording of reactions that take place in the process;	He / She is acquainted with current concepts of chemical literacy; **I G C**	He / he has mastered the skills of chemical expression; **I G C**
Teacher comment:		

Evaluation of student (s)-responsible for individual tasks:
✓ responsible, efficient, critical / self-critical performance of scheduled tasks;
I G C
✓ striving for self-development;
I G C
✓ commitment to the achievement of individual and group goals;
 I G C

Teacher comment:		
Generic competency: The ability to **learn by solving problems**		
student's **Activities / individual Tasks**	student's **Knowledge**	student's **Skills**
Teacher comment:		

2. *Knowing the importance of **raising / leavening agents** as an issue in the confectionery industry;* **I G C**	And / She <u>knows:</u> ✓ the operation of these agents, ✓ the purpose of leavening and baking as baker's operations. **I G C**	And He / She <u>uses:</u> ✓ logical thinking, ✓ with appraisal of the quality and validity of current information to reach an appropriate conclusion; **I G C**
Teacher comment:		
*3. Integration **of current technologies** in the production process of chemical process with the properties of manufactured products used for everyday needs*	He / She <u>recognizes:</u> ✓the chemistry of the current properties of items in daily use as a result of the completion operations of the technological process. He / She <u>provides:</u> problems and opportunities, and the optimal production conditions of a given item. **I G C**	And He/ She <u>knows:</u> ✓ to take careful observation; ✓ to define the main factors for necessary properties of planned products; ✓to clarify the logical cause-effect link between the technological processes and the properties of products of the chemical-processing industry. **I G C**
Teacher comment:		

Evaluation of student(s)-responsible for individual tasks:
✓ responsible, efficient, critical / self-critical performance of scheduled tasks;
I G C
✓ striving for self-development;
I G C
✓commitment to the achievement of individual and group goals; **I G C**

Teacher comment:

Contextual keyword 2:

Soda for experimental enthusiasts

Group 2: Experimenters

Table **11:** Descriptive evaluation of key generic competences for **experimenters**

Generic competences: **1.** The ability to **collect information,** **2.** The ability to **learn by problem solving;**		
student's **Activities / individual Tasks**	student's **Knowledge**	student's **Skills**
1. Identification of components (topical chemicals) from the given cake recipes.	He / She <u>knows the</u> components of a recipe and can identify chemical compounds. **I G C**	He / She <u>follows</u> the instructions for collecting, analysing and organizing information. **I G C**
Teacher comment:		
2a. Determination of chemical formulas in printed resources (chemicals).	He / She is <u>acquainted</u> with the language of chemistry. **I G C**	He / She can <u>evaluate</u> the quality and validity of information. **I G C**
Teacher comment:		
2b. Identification and **laboratory obtaining** *a compound with the*	He / She can <u>describe</u> the Solvay process of soda production.	He / She is <u>planning</u> his/her own work; **I G C**

ability to raise dough (formula Task 2a)	**I G C**	
3.Experimenting with the reaction of the substance with the effect of expanding and raising dough;	He / She can <u>explain</u> chemical reactions of production and demonstration (proof) of carbon dioxide (formula CO_2 as a leavening in confectionery. **I G C**	He / She has <u>acquired</u> the level of experimentation and skill in carrying out current laboratory techniques, while taking into account security issues; **I G C** He / She <u>make inferences</u> about the formation of gas and the choice of current demonstrative reaction. **I G C**

Teacher comment:

Evaluation of student(s)-responsible for individual tasks:

✓ responsible, efficient, critical / self-critical performance of scheduled tasks;
I G C
✓ striving for self-development;
I G C
✓ commitment to the achievement of individual and group goals;
I G C

Teacher comment:

Contextual keyword 3:

Group 3 Consumers

Table 12: Descriptive evaluation of key generic competences for **consumers**

Generic competence 1: The ability to **collect information**		
student's **Activities / individual Tasks**	student's **Knowledge**	student's **Skills**
Asking questions as a method of studying the meaning of the term in everyday use.	✓ He/She knows the importance of the (chemical, food, healing, cleansing) selected consumer items; **I G C**	✓ He/She knows how to use the selected information source; **I G C** ✓ He/She has a critical attitude towards the information collected. **I G C**
Teacher comment:		

Generic competence 2: The ability to **learn by problem solving;**		
student's **Activities / individual Tasks**	student's **Knowledge**	student's **Skills**
Interpretations (theoretically) defined research problem for the needs of responsible consumer.	✓ He /She knows current degree of the problem solving method; **I G C**	✓ He /She Independently formulates and implements the plan necessary for their own work on the problem-solving task; **I G C** ✓ He /She has acquired a strategy for planning problem learning. **I G C**
Teacher comment:		

35

Generic competence **3:** The ability to work **independently and in a team**

 ✓ individual chooses a task and thereby determines their own share in the realization of the role played by a group exercise; **I G C**

 ✓the person responsible for task can <u>explain</u> to members of the same / other groups the objectives of the specific tasks and explain how to solve the latter as part of the overall task (mission); **I G C**

 ✓ the person responsible for the task is able to assess the duration of the individual task and adjust the pace of the work, to suit the time frame of teamwork; **I G C**

Electronic balance KERN AES/AEJ (Photo: Iris Petrovič)

Topic two

I   What is Hidden in Student Laboratory Notebook

*Motto: **No one knows what he can do till he tries**. (Publilius Syrius)*
All Science is formed necessarily of three things:
the series of facts which constitute the science; the ideas which they call forth;
***the words which express them**. (Laurent –Antoine Lavoisier)*

Contents of the Booklet

Summary

This booklet forms an example of didactic material (short title: LabNotebook), which directs pupils / students as experimenters in making personal recordings of their laboratory work. Pupils'/ Students' notes should include a list of current experimental skills and self-evaluation of the implementation of these skills: skills that have been perfectly mastered but also those that could be improved.

The second substantive part of these notes would include the properties of substances faced by students in chemistry class. Thus, these teaching materials would become a student's personal collection of basic information on a variety of chemical materials.

The theoretical part of this booklet comprises the target defined didactic material, based on the following:

(1) A Catalogue of knowledge of the subject "Laboratory and Analytical Technique" (Chemical Technician, for the professional baccalaureate program,

(2) Curriculum for the subject "Chemistry" (general and professional grammar school, for the general and professional baccalaureate).

Specifically defined are topical generic competences as well as specific scientific competences, which can also be achieved in the regular learning process using this didactic material.

The guide for the teacher outlines the work strategy, while the learning guide for the pupil presents one example of a sample articulated LabNote. The attached instruments of evaluation, which highlight current generic or scientific competences (for both teacher and student) also represent a key component of this LabBooklet.

Undeniably, the individual notes in this LabBooklet can serve each pupil in self-monitoring of his (her) progress as an experimenter. Of course, the learning attention paid by each student so systematically cultivated also contributes to the improvement of his (her) response to the new learning as well as new everyday situations.

Introduction

Chemistry is a branch of the exact natural sciences, which are based on the experiment as a basic method of scientific research in the field of these sciences. So, using a research experiment allows one to identify the essence of chemical phenomena and to study the legality and conditions of their course. Therefore, experimental work so constitutes an irreplaceable method - the source of knowledge in the field of chemical education. Thus, by using educational research experiments are all three didactic situations under consideration, the consolidation and verification of learning material can be realized.

The right chemistry classroom should be a space in school for the implementation of experiment-guided discovery and verification of chemical principles and skills as well as for acquiring the skills of experimental work, as in a school laboratory. In the chemistry laboratory the teacher and his laboratory assistant train the pupils /students for independent acquisition of chemical knowledge. This means getting used to focusing on observation of what is happening in the experiment and knowing when to stop based on collected observations and interpretation, as well as independent planning new experiments are all needed for efficient solutions of given pseudo-realistic learning problems.

An integral part of formal school education and training a pupil / student as an experimenter should be pupil / student self-education. This ranges from a planned personal record of laboratory work through self-evaluation of experimental design competences, to a pupil's / student's personal collection of basic information on the various substances with which he himself is now experimenting, or which he encounters every day in the living environment.

In addition to these generic competences (in particular the organization and planning of laboratory work), supported by a common scientific competences (specific to the field of chemistry, biology and physics), it is also important for the pupil / student experimenter to acquire the chemistry competences (subject-specific), which should include skill in safe handling, especially safe experimentation with chemicals. To prepare the pupils / students - experimenters for systematic monitoring of their own development, these competences will serve as instruments for evaluation, as submitted in this booklet. Implementing self-assessment of experimental skills as an example of these competences begins with the creation of individual notes (lab notes). Regular preparation of this Laboratory Notebook will enable the pupils / students to record their own progress in the role of the experimenters.

In the theoretical part of this didactic booklet, we begin from the record of the general objectives / competences as defined in the Slovenian Catalogue of skills in the subject of laboratory and analytical technique, which is designed to be implemented at the level of the vocational Matura in secondary vocational education, more precisely in the profession of Chemical Technician. Of course, the validity of this material also relates to the subject of Chemistry in the general and professional baccalaureate education program for grammar schools at the level of general secondary education in the Republic of Slovenia.

Since both school experiments and experimental work in the laboratory remain the most important components of chemistry teaching, the applied part of this booklet comprises three key components: a didactic guide (for teachers and pupils / students), work instructions for pupils / students (i.e. a learning booklet for pupils / students and pupil / student self-evaluation material. Both types of materials direct pupils / students to the activities that must be performed before, during and after the experiment. Only then will they be able to act either as active observers or as safe contractors, as well as rational planners of laboratory experimental work. So, you can direct the young pupil / student experimenter to self-education as a future responsible member of the community, not only for the selected occupation, but also as a member of the local community or as a socially aware consumer. For successful implementation of the teacher's role "primus inter pares" in the democratic-controlled learning situation, instructions for the teacher are included, describing his strategy for teaching-instructing-directing the work.

Among the many potential causes of accidents in carrying out experiments, these stand out: pupil / student ignorance, clumsiness, frivolity and disobedience. This last should be understood as any arbitrary experimentation performed over the teacher's instructions (ignoring them). So, only pupil / student experimentation under supervision by teachers / laboratory assistants guarantees absolute security. Therefore, the proportion of a pupil's / student's self-education as experimenter in the form of the proposed "lab notebooks" should not be overlooked, but this approach would be encouraged and developed.

Theoretical section

1. The teaching and learning approach: **The Purpose and Objectives of the Program**

Programs:
Secondary Professional Education (SPE) - *Chemical Technician Program*
Secondary General Education (SGE) - *Grammar School*

Subjects (modules):
(1) Laboratory and analytical technique → Vocational Baccalaureate
(2) Chemistry→Matura and Professional Baccalaureate

Objective-Competence activities:
This didactic material for implementation of the experimental work provides pupils / students:

(1) Training in the acquisition of the professional competences, identified in the content section "Laboratory Technique" in the subject "Laboratory and Analytical Technique" in preparation for an experimental project task in a professional baccalaureate (Table 1, section 1); (2) Developing experimental skills and a research approach within the secondary school program of the matura (in Slovenian) and professional baccalaureate (Table 1, section 2).

Table **1: Guidance** objectives and (vocational) competences from the catalogue of skills in the program **Chemical Techniques** in the Grammar School Curriculum

Guidance Objectives	Professional Competences
The content section: **Laboratory Technique**	section 1
Common Competences objectives: ✓ *Independent planning of laboratory work;* ✓ *Knowing the basic technical terminology;* ✓ *Combining of theoretical knowledge with practical;* ✓ *Collecting data, make sense to regulate them and intervene to conclude on the natural laws;* ✓ *Introduction to the methodology of research work;*	
✓ to know and use the chemical laboratory and laboratory	(A) Use laboratory

41

equipment; ✓ to know the basic work techniques, develop skill at and accuracy in working according to the principles of good laboratory practice;	equipment and perform basic operations
✓ to consider the rules for safe operation and use of personal protective equipment; ✓ to deal responsibly and safely with materials in health care; ✓ to provide for the protection of the environment when disposing of chemicals;	(B) Safe handling of chemicals
✓ to know the measurement procedures and measurement devices; ✓ to master chemical stoichiometry and problem solving; ✓ being able to assess the adequacy of the results obtained and determine the size of the error; ✓ to make a report on one's own work.	(C) Preparing a solution
✓ to use professional literature and laboratory manuals; ✓ to acquire basic skills important for quality performance of complex job tasks; ✓ critically evaluate one's own work.	(D) Employing separation and purification of substances
The content section: **Introduction to Safety in Experimental Work**	
✓ to define experimental- research problem and set experimental-research questions, formulate hypotheses and predict possible solutions; ✓ to identify dependent and independent variables and their control; ✓ to connect, compare and critically evaluate the obtained experimental results with the results published in various professional sources, as well as to propose changes, additions and upgrades;	(A) Develop experimental skills and introduce oneself to the methodology of the research approach
✓ to plan a course of safe experimental-research work, which includes the search for suitable experiments and consequently knowledge of basic laboratory equipment and techniques, as well as strictly to observe the rules of safe experimentation.	(B) Observes the principles and rules of chemical safety

2. The teaching and learning approach: **Competence Activities**

a) Generic competences

One of the optimal generic competences that can be achieved with the teaching and learning of the proposed topics using the didactic material (addressed as "What Is Hidden in Pupil / Student Laboratory Notebooks", shortly Lab Notebook) is "adjusting to new situations"; its application is presented in Table 2.

Table **2**: Application of the **generic competency** "adjustment to new situations"

Competence activities	Pupil's / Student's knowledge	Pupil's / Student's skills	Pupil's / Student's relations
➡ guidance notes of laboratory work ➡ collecting of descriptions of the properties of the substances;	✓ understanding the theoretical bases;	✓ responsiveness, ✓ focussing;	✓ to follow instructions accurately and attend to advice and recommendations (from teachers and technical staff);
The teacher's notes:			

Other generic competences by student-experimenter develop as follows:

✓ abilities to: **(1)** gather information, **(2)** perform independent work and teamwork, **(3)** synthesize findings, **(4)** analyse and organize information;
✓ qualifications for: (5) the organization and planning of work, **(6)** verbal and written communication, **(7)** ensuring chemical safety.

b) Natural (subject-specific) competences

Table **3**: Taxonomic evaluation of **subject-specific competences** in the didactic booklet Lab Notebook

Subject-specific *capabilities* available for learning using the Lab Booklet; *taxonomic level of the target area / Bloom or Marzano*
✓ compliance with the basic rules of safe laboratory work; *work with sources / Marzano* ✓ self-reflection and self-assessment in conducted experimental-research learning; *evaluation / Bloom* ✓ identification errors in their own experimental work; *analysis of errors / Marzano* ✓ optimal methods and techniques in implementation of laboratorial work; *understanding and application / Bloom*

3. Prior knowledge

Pupil's / Student's current pre-knowledge represents the theoretical and skill preparation for carrying out the forms of experimental work, namely the experimental work of learners or just monitoring experiments (so-called demonstration experiment).

4. Glossary of curricular and chemical terms

Competences: a term which is in general defined as a combination of knowledge, skills and attitudes appropriate to the context. There is well established distinction between the key, generic, specific (specific) and subject-specific competences.

In the implementation of laboratory work, a pupil-experimenter obtains (develops) a whole range of generic competencies (14 per Mayer, 1991) of which we highlight the following seven skills: (1) analysis of the literature and organization of information, (2) the transfer of theory into practice, (3) organization and planning of the work, (4) interpretation, (5) learning and problem solving, (6) independent work and teamwork, (7) safety assurance.

The following explanation of examples of chemical terms for some of the most commonly used laboratory techniques:

Weighing one of basic methods for measuring quantities of chemicals:

✓solid are weight using one kind of containers (glass, plastic, aluminium);

✓liquid are weighed directly into the reaction flask.

Filtration: separation process of suspended solids from a solution by percolation through a filter.

Distillation: a process of heating a liquid to the boiling point, condensing the heated vapour by cooling it and then returning only a portion or none of the condensed vapours to the distillation flask. This technique can be used for separating two or more substances on the basis of their differences in vapour pressure.

Precipitation: a chemical reaction which results in the solution of insoluble substances.

Neutralization: the process in which an acid reacts with a base to form a salt and water.

In generally, salt as the compound is formed in replacing the hydrogen of the acid by a metal or other positive ions.

Titration: a procedure of volumetric analysis method in which a volume of one reagent (of a known concentration) is added to a known volume of another reagent (of unknown concentration) slowly until an end point is reached (using indicator, such as phenolphthalein as an acid-base indicator) and the unknown concentration can be calculated.

5. Learning resources

Asselborn, W., Demuth, R. (1992). Chemieunterricht ohne Entsorgungsprobleme. Schroedel Verlag GmbH, Hannover.

Atkins, P.V. , Frazer, M.J., Clugston, M.J., Jones, R.A.Y. (1988). Chemistry. Principles and Aplications. Longman Group UK Limited, London.

CVK Chemie für Realschulen. Ein neues Arbeits-und Informationsbuch (1986). Cornelsen-Velhagen Klasing GmbH Co Verlag für Lehrmedien Kg Berlin.

Gallagher, R.M., Ingram, P. (1991). Co-ordinated Science - Chemistry. Oxford University Press.

Hagenauer, Böhm, Jarisch, Markl, Pribas , Zadrazil (1994). Chemie aktuell 1, Arbeitsheft. Naturwissenschaftliche Verlagsellschaft m.b.H, Salzburg.

Hagenauer, Jarisch, Markl, Pribas, Zadrazil (1990). Chemie aktuell 1, Naturwissenschaftliche Verlagsellschaft m.b.H, Salzburg.

Lister,P. (2008). Work Safer=Work better. Edizioni Universita di Triesta, Triesta.

MacKenzie, Charles A. (1963). Experimental Organic Chemistry. Prentice-Hall.USA.

Mendoza E.E., Religioso, T.F. (1996). Chemistry. Phoenix Publishing House. Quezon City.

Meyendorff, G. (1977). Chemische Schullerexperimente. Volk und Wissen Volkseigner Verlag, Berlin.

Pavia, D.L. , Lampman, G.M., Kriz, G.S., Engel, R.G. (2011). A Small-Scale Approach to Organic Laboratory Techniques, 3-rd edition. Brooks /Cole, Cengage Learning, Transcontinental. (useful for teachers)

Safety in the Chemistry Laboratory a practical guide for teachers. Ed. Stepnowski Piotr and Wilamowski Jaroslaw. Fundacija Rozwoju Uniwersytetu Gdanskiego, Gdansk.

Schüller duden, Chemie (2001). Bibliographisches Institut & F.A. BrockhausAG, Mannheim.

Learning of Chemistry our secondary students as experimenters (Photo: Darinka Sikošek)

Applied section

The contents of the applied part of this didactic booklet comprise the following components:

✓ Instructions to teachers for the purpose of didactic planning, i.e. Didactic Proposal (A1) and advance implementation of this defined experimental work, i.e. Implementing Guidance (A2)

✓ Instructions to pupils, i.e. Learning guide for pupils (B);

✓ Didactic material in the form of a template for articulating this learning-worksheet (C);

✓ Another type of didactic material for pupils' self-evaluation of their own laboratory work (D).

A. Instructions to teacher

1. Didactic proposal

Suggestion for optimal teaching methods and social learning forms of work, which should be used in the implementation of students' activities planned in the framework of didactic material Lab Notebook is evident from Table 4.

Table 4: **Didactic proposal suggested teaching methods and social learning forms**

Pupil's Activities	Teaching Methods	Forms of Social Learning
✓ practice recording of laboratory work; ✓ collecting descriptions of the properties of materials;	✓ planning and performance of experimental work *(guiding method)* ✓ working with text *(supporting method)*	✓ *individual work* ✓ *individualized teamwork*

2. Implementing advice

1. The teacher distributes the individualized worksheets to each of the pupils / students, who read these independently.

2. Teachers use the attached example of a record (sample) for the laboratory notebook (in the form of a transparency) so that pupils / students can gain an idea of the format and content of such a record (see section B. The learning guide for pupils; slide 1).

3. Before the next series laboratory exercises, each pupil / student has to prepare his /her notes (according to the instruction hand-out), which the teacher consistently verifies.

4. Pupils create records in their personal Lab notebooks during the implementation of the entire lab work program or during a pre-determined time-content period.

5. After finished period of working with Lab notebook (i.e. the conduct of individual notes) pupils /students individually (when given a self-evaluation sheet) self-evaluate the quality of their own work. Each pupil /student is obliged to submit a personally completed self-evaluation sheet to his/her teacher.

6. The teacher reviews these self-evaluation sheets and assesses the pupil's / student's acquired skills, using the attached evaluation instruments.

B. Learning guide for pupils / students

Introductory note: Slide 1, below represents exemplary notes for the "Lab notebook", the laboratory-exercise "Determining the density of liquids with a pycnometer". "So you can get an idea of the form and content of the notes, which will help you in completing individual worksheet."

Transparency **1**: Example of a LabNotebook record

Content keyword: <u>Laboratory view of the world of substances</u>

Experimental keyword: **Determining the density of liquids using a pycnometer**

1. Theoretical basis

In general, the density is defined as the ratio between weight and volume of a body. The density of a substance is characterized by a constant, which depends on the temperature, as well as the substance.

2. Important factors in experimental performance

For successful experimentation, it is necessary to take into account at least the following three factors:

✓ After each thermostating, the pycnometer should be carefully dried in a laboratory drier;

✓ Between each weighing the pycnometer should be well purified, which means that it is washed several times with distilled water;

✓ The sample in the pycnometer must be mixed (by inverting the container) until the temperature reading on the thermometer has a constant value.

3. Safety Measures

In order to maintain chemical safety, the following steps must be taken:

✓ Check the settings of the apparatus in use;

✓ Before weighing, ensure the purity of the scales, to exclude the presence of factors that could affect the result (wind, vibrations etc...);

✓ Check if the settings of vacuum dryer are correct.

4. Teacher and laboratory assistant advice

Be careful to follow this recommendation:

✓ The stopper of the apparatus serves as an expansion chamber; therefore, it is necessary to handle it carefully, because the stopper as precision-made ground glass apparatus.

5. References

For better understanding of the planned work, the following textbook source is recommended:

Lister,P. (2008). Work Safer=Work better. Edizioni Universita di Triesta, Triesta.

Meyendorff, G. (1977). Chemische Schullerexperimente. Volk und Wissen Volkseigner Verlag, Berlin.

Szafran, Z., Pike, R.M., Singh, M.M. (1991). Microscale Inorganic Chemistry. A Comprehensive Laboratory Experience. John Wiley & Sons, USA.

6. Descriptions of chemicals used

Example: *A solution of sodium chloride NaCl (aq)*

The solution of each substance has the characteristic physical and chemical properties. If water is added to solid sodium chloride, then an aqueous solution of sodium chloride NaCl (aq) is formed. Sodium and chloride ions, which are associated with a strong ionic bond of the crystal structure in the process of dissolution, are mixed with the water molecules, but do not react with them chemically.

A large quantity of sodium chloride is dissolved in seawater. We all encounter this chemical daily, as a nutritional product. As is known, the NaCl is a basic spice (i.e. cooking salt), which is necessary to the human organism. Blood contains 0.9% of this chemical, but in body fluids, there is from 150 to 300 g of salt. Our daily requirement of NaCl is 10 to 15 g, but an excess amount of salt is detrimental. For medical needs, there is well known physiological saline which is 0.9% sodium chloride solution in water.

C. Didactic teaching material ➡ **Learning-Worksheet**

Introductory note: Using the attached individual learning-worksheet, the pupil (you) provides answers (detailed instructions) about how the pupil's (your) LabNotebook is used to design and "write up" personal notes about the pupil's (your) laboratory work.

Individual learning-worksheet

Title of LabNotebook:

(My) Laboratory view of the world of substances

Question 1: **What is my task?**
Suggestion 1: *Preparing (designing and creating notes) a personal laboratory notebook (LabNotebook).*

Question 2: **How should I create a LabNotebook?**
Suggestion 2: *Aim for an imaginatively designed notebook (Be creative). On the first page write (even the sign) the address:*
(My) Laboratory view of the world of matter.

Question 3: **And how will I use this notebook?**
Suggestion 3: LabNotebook offers to you particularly useful possibilities as they are:

(1) In performing laboratory exercises, you can into write down answers to these questions:
Which chemicals do you know? What are their characteristics (properties)?, How can I safely handle them?
(2) After completion of each laboratory exercise, you can write down these answers:
How you did experiment; What you learned (according to the instructions for tasks 1, 2); How the experiment could be improved. (Don't forget this part.)

Question 4: **What will I achieve with all this recording?**
Suggestion 4: The notes (records) in your LabNotebook represent a handy "lexicon", which provides you with an instant comprehensive review of your own competences in chemistry knowledge and laboratory skills.

Do not forget, it is quite common to be unable to remember something that we already knew or could do. In such case these notes (records) in this LabNotebook will not only help to evoke the necessary information- i.e. what and how to do it-, but will serve as a means to discourage actions, such as repeating old mistakes when preparing, for example, final project assignment for your professional baccalaureate.

Now let's move from questions to imperatives, thus testing the Latin proverb:
Vox audita perit, littera scripta manet! *(Heard the word is lost, written remains!)*
Carefully read both Tasks 1 and 2, and give serious thought and attention to solving them.

Task 1: Laboratory work
Note: After completing the experimental work, make a record (using the questions below) for your LabNotebook, regarding each experiment as separately implemented.
  1. Which theoretical knowledge is needed to understand the experiment (brief explanation).
  2. During the execution of the experiment, which agents require special attention? (Please be specific)
  3. What care in handling had to be taken for secure implementation of the planned experiment? (Brief explanation)
  4. What advice did you need from a teacher or laboratory assistant? (Brief explanation)
  5. What literature did you use when drawing conclusions? (Please be specific)

Task 2: A look into the world of the substance
Note: During the school year, diligently record in a LabNotebook descriptions of substances (chemicals), from these that are used in the school laboratory, to those that we hear about through mass media (especially television or radio), as well as those that you can see in shops or in specialized stores.
Special care should apply to the properties of those chemicals with which you are yourself experimenting. (You should make extra effort to create a meticulous record)

The present evaluation of instruments is designed for use by both teacher and student.

The evaluation of the level of competence covers the three generic competences
✓ability to experiment safely, ✓ability to draw conclusions, ✓ability to collect information.

Also taken into account is the principle of differentiation and individualization, which is evident from the used three-level approach (see Table 4, for teacher use only).

The pupil's self-identified achievement of generic competences is evident from the pupil's completed self-evaluation sheet, which is reviewed by the teacher, who identifies the level of the individual generic competencies achieved, determines progress and develops an evaluation (also a mark) for the pupil as an experimenter.

Table **5**: Level descriptions of selected achievements in the field of **generic competences**

Pupil activity	Level 1	Level 2	Level 3
Generic competency: **The ability to gather information**			
Planning and conducting the experimental approach	Follows the instructions for collecting, analysing and organizing information.	Can obtain information from different sources	Assesses the quality and validity of information

Generic competency: **The ability to synthesize conclusions**			
Designing a record of laboratory work	Identifies the desired results and procedures.	Explains the desired results and procedures and their interaction.	Determines the main factors that affect procedures and results.
Generic competency: The **ability to perform experiments safely**			
Experimentation	Is aware of the dangers and knows the rules of safe work.	Prepares a risk assessment then ensures safe execution.	Prepares a risk assessment and ensures safe execution, as well as helping their classmates with the safe version.

Pupil Self-Evaluation sheet

When using the self-evaluation sheet, pupils evaluate themselves as experimenters in the first task, while the second task concerns their self-evaluation as competent experts in chemical substances.

Self-Evaluation Task **1: Laboratory work**

Write down the theoretical knowledge that you've acquired in the implementation of the experimental work. (Use a Labnote)

Number each record. It is useful and therefore recommended to complete all records using professional literature.

Example:

1. The density of solids and liquids can be measured with a pycnometer. In doing so, we must ensure that, after each thermostate, the device is carefully dried in the labdrier. The pycnometer must be carefully rinsed with distilled water before each individual weighing

Self-Evaluation Task **2: A look into the world of the substance**

Task 1: In my LabNotebook I have a collection of descriptions of the properties of many materials:

(a) less than ten descriptions, (b) 10 to 30 descriptions, (c) more than thirty descriptions

Write down the names of substances and their properties which are written in a note by task 2.

Task 2: A look into the world of each collected substance. As a useful technique it is recommended that even these records be completed with notes from the professional literature.

View in the LabClassroom of our-students-teachers of chemistry (Photo: Darinka Sikošek)

Subject index

A

abilities, 6, 10-11, 14, 43
ability
    to collect information, 30, 33, 35, 53
    to learn and to solve problems, 9
    to perform experiments safely, 54
    to work independently and in a team, 30, 36
    to synthesize conclusions, 54
accidents, 40
active observers, 40
additional literature for teachers, 6
ammonia, 15-16, 26
antacids, 29
aqueous solutions, 7-8, 15

B

background knowledge, 6
baking soda, 15, 22, 24, 29
baking soda (sodium hydrogen carbonate), 29
Base of Biotechnology for Nutrition, 8
basic chemical educational standard, 6
bicarbonate of soda, 3, 5, 7, 12, 28

C

chemical, 2-10, 12-15, 19, 21-24, 26-27, 31-35, 37-45, 49-52, 54
chemistry, 2-5, 10, 15-17, 22, 32-33, 38-39, 40-41, 45-46, 55
competences, 3, 6-7, 9-10, 12, 20, 30, 33, 35, 38- 41, 43- 44, 51
    activities, 9
    targets (aims), 10
considering didactic principles, 11
consumers, 3, 6, 12-13, 19, 25, 28-29, 35
contextual password set, 9, 10
curriculum for the subject "Chemistry", 38

D

density, 48-49, 54
description of strategy, 20
dictionary of chemical terms, 6
didactic
    booklet, 40, 44, 47
    guide, 40
    material, 6-7, 9, 11, 14, 38, 41, 43, 47
    principle of differentiation, 6
    proposal, 7, 18
distillation, 45

E

educational research experiments, 41
evaluation
    instrumentation, 29
    material, 16, 37, 40, 53
examples of activities, *methods*, 8
experimental
skills, 40-41, 43-44

56